University of Liverpool, Liverpool Marine Biology Committee

L.M.B.C. Memoirs on Typical British Marine Plants and Animals

Volume 1: Ascidia

University of Liverpool, Liverpool Marine Biology Committee

L.M.B.C. Memoirs on Typical British Marine Plants and Animals
Volume 1: Ascidia

ISBN/EAN: 9783337237615

Printed in Europe, USA, Canada, Australia, Japan

Cover: Foto ©berggeist007 / pixelio.de

More available books at **www.hansebooks.com**

Liverpool Marine Biology Committee.

L.M.B.C. MEMOIRS
ON TYPICAL BRITISH MARINE PLANTS & ANIMALS
EDITED BY W. A. HERDMAN, D.Sc., F.R.S.

I.

ASCIDIA

BY

W. A. HERDMAN, D.Sc., F.R.S.

Professor of Natural History in University College, Liverpool
Hon. Director of the Port Erin
Biological Station

(With 5 Plates)

PRICE EIGHTEENPENCE

LIVERPOOL
T. DOBB & CO., PRINTERS, 229 BROWNLOW HILL
OCTOBER, 1899

I.

ASCIDIA

NOTICE.

THE Committee desire to intimate that no copies of these
Memoirs will be presented or exchanged, as the prices
have been fixed so low that most of the copies will have
to be sold to meet the cost of production.

The Memoirs may be obtained, post free at the nett prices
stated, from the Hon. Treasurer, Mr. I. C. THOMPSON, 53,
Croxteth Road, Liverpool; Prof. HERDMAN, University
College, Liverpool; or the CURATOR, Biological Station,
Port Erin, Isle of Man.

Memoir I.   Ascidia—now ready. 60 pp. and five plates,
               price 1/6.
  ,,   II.   Cardium—in a few weeks, price 2/-.
  ,, III.   Echinus—at Christmas, price 1/-.

Liverpool Marine Biology Committee.

L.M.B.C. MEMOIRS

ON *TYPICAL BRITISH MARINE PLANTS & ANIMALS*

EDITED BY W. A. HERDMAN, D.Sc., F.R.S.

I.

ASCIDIA

BY

W. A. HERDMAN, D.Sc., F.R.S.

Professor of Natural History in University College, Liverpool
Hon. Director of the Port Erin
Biological Station

(With 5 Plates)

LIVERPOOL
T. DOBB & CO., PRINTERS, 229 BROWNLOW HILL
OCTOBER, 1899

EDITOR'S PREFACE.

THE Liverpool Marine Biology Committee was constituted in 1886, with the object of investigating the Fauna and Flora of the Irish Sea.

The dredging, trawling, and other collecting expeditions organised by the Committee have been carried on intermittently since that time, and a considerable amount of material, both published and unpublished, has been accumulated. Thirteen Annual Reports of the Committee and four volumes dealing with the "Fauna and Flora" have been issued. At an early stage of the investigations it became evident that a Biological Station or Laboratory on the sea-shore nearer the usual collecting grounds than Liverpool would be a material assistance in the work. Consequently the Committee, in 1887, established the Puffin Island Biological Station on the North Coast of Anglesey, and later on, in 1892, moved to the more commodious and convenient Station at Port Erin in the centre of the rich collecting grounds of the south end of the Isle of Man.

In our twelve years experience of a Biological Station (five years at Puffin Island and seven at Port Erin), where College students and young amateurs formed a large proportion of the workers, the want has been constantly felt of a series of detailed descriptions of the structure of certain common typical animals and plants, chosen as representatives of their groups, and dealt with by specialists. The same want has probably been felt in other similar institutions and in many College laboratories.

The objects of our Committee and of the workers at our Biological Station have hitherto been chiefly faunistic and speciographic. The work must necessarily be so at first when opening up a new district. Some of our workers have published papers on morphological points, or on embryology and observations on life-histories and habits; but the majority of the papers in our volumes on the "Fauna and Flora of Liverpool Bay" have been, as was intended from the first, occupied with the names and characteristics and distribution of the many different kinds of marine plants and animals in our district. And this faunistic work will still go on. It is far from finished, and the Committee hope in the future to add greatly to the records of the Fauna and Flora. But the papers in the present series are quite distinct from these previous publications in name, in treatment, and in purpose. They will be called the "L.M.B.C. Memoirs," each will treat of one type, and they will be issued separately as they are ready, and will be obtainable Memoir by Memoir as they appear, or later bound up in convenient volumes. It is hoped that such a series of special studies, written by those who are thoroughly familiar with the forms of which they treat, will be found of value by students of Biology in our laboratories and in Marine Stations, and will be welcomed by many others working privately at Marine Natural History.

It is proposed that the forms selected should, as far as possible, be common L.M.B.C. (Irish Sea) animals and plants, of which no adequate account already exists in any text-book. Probably most of the specialists who have taken part in the L.M.B.C. work in the past, will prepare accounts of one or more representatives of their groups. The following have already promised their services, and in some cases the Memoir is already far advanced. The first three Memoirs will be issued before the end of 1899, and others will follow it is hoped in rapid succession.

Finally, I desire to acknowledge that a welcome donation of £100 from Mr. F. H. Gossage of Woolton has met the expense of preparing the plates in illustration of the first few Memoirs, and so has enabled the Committee to commence the publication of the series sooner than would otherwise have been possible.

W. A. HERDMAN.

University College, Liverpool,
    October, 1899.

"The ancestor remote of Man,
  Says Darwin, is th' Ascidian."
                    Anon. Ballad.

"For Man was once a leather bottél."
                    Old Song.

L.M.B.C. MEMOIRS.

No. I. ASCIDIA.

Professor W. A. HERDMAN, D.Sc., F.R.S.

ALTHOUGH the Ascidian has been much talked of and
written about during the last quarter of a century, comparatively few people, beyond the circle of professional
biologists, have an accurate idea of what the animal is like
in appearance and structure, or have more than a vague
notion as to what the popular impression of its relationship to higher animals is based upon.

The adult or fully-developed Ascidian is a very remarkable animal, and presents numerous interesting problems
for the biologist to investigate ; but its reputation is due,
not to any of these, but to certain changes which the
young animal undergoes in its development from the egg ;
changes so remarkable and interesting that when the more
important of them were discovered by Kowalevsky, some
thirty years ago, they gave rise to the belief that Ascidians
were a group of animals holding a position between the
Vertebrata and the Invertebrata, and indicated a possible
line along which the former might have been evolved from
the latter. The opinions of scientific men on this point
have undergone a certain amount of modification during

late years, and Ascidians are now regarded as the degenerate descendants of a very lowly-developed group of the early Vertebrata (or more correctly, Chordata).

But before we are in a position to understand this important matter, it is well to have some knowledge of the ascidian structure at several different stages in the life-history, and we shall commence with the last stage of all—the full-grown, or adult condition.

MODE OF OCCURRENCE.

Ascidians are all marine, and they have been found in all seas, from tropical to polar regions, and at depths varying from near high-water mark down to three and a quarter miles or so. Round most of our own coasts they are common, and some kinds are familiar enough, under the name of " sea-squirts," to many sea-side visitors, as being amongst the inhabitants of rock-pools which can be collected and kept in aquaria. Such forms are usually found as dome-shaped bodies of a dull red colour, adhering to the rock or sea-weed, and having two small openings on their upper surface from which, when touched, they emit delicate jets of sea-water with considerable force—thus establishing their claim to the title of " sea-squirts "; and their resemblance to double-necked leather bottles, whence the name Ascidian (from the Greek ἀσκὸς).

Others again form flat gelatinous expansions attached to sea-weeds or stones, and symmetrically marked with bright spots of colour in the form of circles, meandering lines, or starlike patterns. These are really colonies in which each spot of colour or ray of the star represents an ascidiozooid or member of the colony equivalent to the whole animal in the case of the solitary Ascidian.

By dredging around the coast, outside low-water mark, many other kinds of Ascidians are found, of diverse forms and often most gorgeous colouring. One of the commonest species of our seas, *Ascidia virginea*, O. F. Müller, has the form of a short and somewhat irregular cylinder, with rounded ends, one of which is attached to a stone or dead shell, or some other object at the bottom, while the other end is directed upwards into the water, and bears two short projections, each terminated by an opening (see fig. 1, Pl. II.). This species is of a dull grey colour; it is usually found in from five to twenty fathoms of water, and is in some places so abundant that the naturalist's dredge may come up absolutely filled with it. An average size is an inch and a half in length, an inch in breadth, and half an inch in thickness; but in a dredgeful one usually finds all sizes, from a quarter of an inch to two inches in length. Our largest British species, *Ascidia mentula*, O. F. Müller (see Pl. I.), measures from three to six inches in length, and is usually found on a muddy bottom in from ten to fifty fathoms. Either of these two species of *Ascidia* will serve very well as the type of a simple ascidian, and the following description in nearly all details applies to both.

External Appearance.

The two openings at the upper end, although they appear very similar at first sight, are really different, and can readily be distinguished with a little practice. One of them (*Br.* in fig. 1, Pl. II.), is higher, or more nearly terminal in position than the other (*At.*), which may be placed some way down one edge; the former is the Branchial, and the latter the Atrial aperture. A close examination will show that the margin of the branchial aperture is cleft into eight small projections, or lobes; while the atrial aperture is bounded by six lobes only (see also Pl. I.—as an individual peculiarity this specimen has a seventh small lobe at its atrial aperture). Consequently, we can distinguish the apertures both by their position and by the number of their lobes; and this is a very important matter, since it enables us to recognise the different regions in the body of the Ascidian; and also because the exact number of lobes is a characteristic of certain families and genera of Ascidians, *e.g.*, *Molgula* has six branchial and four atrial lobes, while *Ascidia* has eight branchial and six atrial. Plate I. shows the exact shape of the lobes and the apertures in *Ascidia mentula*.

The branchial aperture (*Br.*) indicates the anterior end of the body, the region which corresponds to the head end of a man, a dog, or a fish; and, consequently, the opposite part, which is attached to the stone, is the posterior end. Then, the atrial aperture (*At.*) is invariably placed to the dorsal side of the branchial; hence, that side of the body near which the atrial aperture is situated corresponds to the back of the man, dog, or fish; while the opposite side is, of course, ventral.

Now that it is known which is the anterior end, and which is the dorsal surface, or back, it is easy, by placing the Ascidian alongside oneself, and comparing the parts, to determine which is the right-hand side and which is the left. In Pl. I. and in Pl. II., figs. 1 and 4, it is the right side of the specimens which is shown. The area of attachment at the posterior end frequently involves a certain amount of the left side.

THE TEST.

The outside of the Ascidian—all that is visible, unless one looks into the expanded apertures—is formed by a stout, gristly, translucent layer called the Test or Tunic (hence "Tunicata," the name of the group), which is notable amongst animal structures for containing "tunicine," a substance which appears to be identical in composition, and in behaviour under various treatments, with "cellulose"—a characteristically vegetable substance, entering largely into the composition of plants. The test is a protective layer, usually about a quarter of an inch in thickness, and may be considered as an exo-skeleton. It is the only tissue of a skeletal nature which the adult Ascidian possesses.

In shape the test is an oblong sac, pierced only by the branchial and atrial apertures, and forming, at its posterior end and left side, the place of attachment by which the Ascidian adheres to the rock (Pl. II., figs. 1 and 4). At this point it may become greatly thickened and expanded to form a margin, or may even grow out in the form of a short stalk, raising the body above the surrounding surface. Stones, sea-weeds, dead shells, and remains of other animals may be overgrown by the test and incorporated in its substance; many sessile animals and various kinds of sea-weeds may become attached to its outside; and some parasitic Amphipod Crustaceans (*Tritæta gibbosa*) and Lamellibranch Molluscs (*Modiolaria marmorata*) may inhabit cavities excavated in its thickness. Microscopic parasitic Algæ may also be present amongst the cells in the interior of the test, and help in giving the animal its colour.

The test is cartilaginous in appearance and consistence, and to some extent in structure, as it consists of a clear or slightly fibrillated matrix in which are imbedded many cells. As the test is morphologically a cuticle, being at first a secretion on the outer surface of the ectoderm (Pl. II., fig. 5, *ec.*), the cells it contains have immigrated to it from the body; and it has been shown by Kowalevsky and others that many of these are mesodermal cells or wandering amœbocytes which have passed through the ectoderm. In the test they reproduce freely and secrete large quantities of the tunicine matrix.

Many of the cells in the test remain small and simple, as the rounded, fusiform, or stellate test cells (fig. 5, *t.c.*). Some become larger, much branched, elongate into fibres, or degenerate into globular pigment cells; others may store up reserve products; while others again are converted into the large vacuolated "bladder-cells" which, in the outer part of the test of *Ascidia mentula*, form a well-marked vesicular layer (Pl. III., fig. 9, *bl.*). In this the structureless matrix contains innumerable closely packed spherical vacuoles, each with a thin peripheral film of protoplasm and a parietal nucleus. These bladder cells measure from 0·10 to 0·15 mm. in diameter. Some of them show more than one nucleus, and may be formed by the fusion of several cells. Some of them at least appear to be derived from the "testa-cells" of the embryo, and are thus descendants of cells belonging to the follicle which surrounds the ovum.

The test also becomes organised by the growth into it of the so-called "vessels." These are out-growths of the mesodermal body-wall, covered by ectoderm, and containing prolongations of blood channels from the connective tissue of the body-wall. Plate II., fig. 5, shows such an out-growth, and exhibits the general relations of test (cuticle) ectoderm, and mesoderm. It also explains how it is that the blood channel being pushed out as a loop gives rise to the double or paired vessels seen branching through the test (Pl. III., fig. 9). The two vessels of a pair are one blood channel imperfectly divided by a connective tissue septum. The blood courses out along one side, round the communication in a terminal knob at the end and back down the other side. The terminal knobs are very numerous and form a marked feature in the outer layer of the test (fig. 9, *t.k.*); in some Ascidians they probably form an accessory organ of respiration. In the adult *Ascidia* the vessels enter the test by a double trunk ventrally, near the posterior end of the left side (see Pl. III., fig. 10).

ECTODERM AND BODY-WALL.

Inside the test, and lying between its inner surface and the muscular body-wall ("mantle"), is a very delicate membrane, formed entirely of small cubical or more flattened cells (Pl. II., fig, 5, *ec.*), with delicate "secreting processes" projecting into the test. This is the Ectoderm, the outermost complete layer of cells in the body; and the test lying over it has been produced as a sort of gelatinous exudation upon the surface of the ectodermal cells, aided by the numerous mesoderm cells which have migrated into it, and which give it the appearance of a connective tissue. Besides the ordinary epithelial cells, a few gland cells and pigment cells may be found in the ectoderm. The ectoderm is turned in for a short distance at the branchial aperture (mouth) and atrial aperture (cloacal), as a short stomodæum and proctodæum, lined in each case by a delicate prolongation of the test (Pl. II., fig. 4, *Br.* and *At.*).

Inside the ectoderm lies a thicker layer, the so-called " Mantle " or body-wall (parietal mesoderm), containing a large number of muscles, which run some along the length of the body, and others across it, so that they form a rude interlacing net-work (Pl. II., fig. 4, *m.*), which is much more strongly developed on the right than on the left side of the body. The shape of the body can be changed, to a slight extent, by means of these muscles in the body-wall. When the Ascidian is killed by re-agents it is generally found that the muscles have contracted and drawn the mantle away from the test, so that these layers are no longer in continuity except on the branchial and atrial siphons, and at the posterior end where the vessels enter the test.

The body-wall (Pl. II., fig. 5) is largely formed of connective tissues, both homogeneous and fibrous, with cells, blood sinuses or lacunæ, nerves, and the many muscle bundles, large and small, formed of long, fusiform, non-striped fibres. The largest muscle bundles are found about the centre of the right side, where they may be 0·5 mm. in thickness. This part in the living *Ascidia* is often brilliantly pigmented — red, yellow, and opaque white—the coloured cells being exactly like those found in the blood. The connective tissue cells or corpuscles are fusiform, stellate, or amœboid, and may become pigmented or vacuolated, like the similar cells of the test.

At the anterior end the body-wall is prolonged outwards to form the two well-marked siphons, or short wide tubes, which lead in from the branchial and atrial apertures. These are surrounded by strong sphincter muscles (Pl. II., figs. 2, 6, and 7, *sph.*). Inside the body-wall lies the large cavity called the Atrium, or the peribranchial cavity, which communicates with the exterior through the atrial aperture, and serves to convey away the water which has been used in respiration (see Pl. II., figs. 2 and 4, *p.br.*).

The ectodermal lining of the atrial or peribranchial cavity has been called by some French writers the third tunic—the first being the test and the second the mantle. The cavity of the atrium is traversed by numerous vascular strands of mesoderm, called connectives (Pl. II., fig. 2, *con.*), passing from the body-wall inwards to the branchial sac.

Figure 6 on Pl. II. shows the relations of ectoderm (with test over it), mesoderm, and endoderm in a section through the antero-dorsal part of the body, where *t*. indicates the test, and *sph*. the sphincter of the branchial aperture. The cavity marked *p. br.* is a portion of the atrial cavity lined by ectoderm, and not to be confounded with a coelom or body-cavity. The absence of a true coelom in the mesoderm will be noticed in this and the other figures (*e.g.* fig. 4); and yet the Tunicata are Coelomata—although it is very doubtful whether an enterocoele is ever formed, as has been described by E. van Beneden and Julin in the development of some. The primitive coelom is, however, largely suppressed during development, and is only represented in the adult by the pericardium and small cavities in the renal and reproductive organs and ducts, as will be shown further on.

CAVITIES OF THE BODY.

The following list of the cavities present in the body of the adult *Ascidia* may be useful at this point :—

1. The alimentary canal, including the branchial sac. This is derived from the archenteron of the embryo, is lined throughout by endoderm, and the system of cavities of the intestinal gland is to be regarded merely as an outgrowth from the alimentary canal.

2. The peribranchial (atrial) cavity, derived from two lateral ectodermal invaginations which join dorsally to form the cloaca and open to the exterior by the atrial aperture.

3. The original embryonic segmentation cavity (blastocoele) remains, where not obliterated by the development of the mesodermal connective tissue, as the irregular system of blood spaces, with its outgrowths in test and branchial sac. The heart is only a special part of this cavity which has differentiated muscular walls.

4. The pericardium and epicardium originate as outgrowths from the archenteron. They may therefore be regarded as cœlomic spaces. The pericardium becomes completely closed off and separated from the alimentary canal. The epicardium may form paired tubes of great length, and may remain permanently connected with the branchial sac.

5. The cavities of the renal vesicles and of the gonads and ducts are spaces formed in the mesoblast. They have been variously interpreted:—

(a) As of the same nature as the blood spaces (blastocœlic), or

(b) As formed by a splitting of the mesoblast (cœlomic).

6. The cavity of the neural gland and its duct opening at the dorsal tubercle is derived from the primitive dorsal neural tube of the embryo, and so may be regarded as a part of the lumen of the cerebro-spinal nervous system.

BRANCHIAL SIPHON AND TENTACLES.

The branchial aperture opens into a large cavity, the Branchial Sac (Pl. II., fig. 4, br.s.), which is merely the anterior portion of the alimentary canal, corresponding to the pharynx or back of the mouth in man, enlarged and greatly modified so as to act as a breathing organ, or branchia—whence its name—in addition to performing other important functions.

The branchial aperture itself (Br.) is thus the mouth of the Ascidian, and the siphon is therefore the commencement of the alimentary canal. Its inner surface is lined for a short distance by a prolongation of the test, and where this stops, at about the line of junction of the ectoderm of the stomodœum with the endoderm of the mesenteron, a circle of delicate hair-like Tentacles (Pl. II., fig. 4, *tn.*) projects into the mouth cavity and forms a sensitive sieve or strainer, through which all the sea-water, and its contents, drawn into the branchial sac has to pass.

The tentacles are simple and tapering in *Ascidia*, but in many other Ascidians they are compound and may be very elaborately branched (*e.g.*, in *Cynthia* and *Molgula*). They are in many cases of different sizes arranged alternately or with some marked symmetry. In *Ascidia mentula* there are usually from 70 to nearly 100 tentacles, of which one-third, say 20 to 30, are much larger than the rest. The rule is for two, occasionally three, much smaller tentacles to be placed between each pair of larger ones visible to the eye. Fig. 7 on Pl. II. shows, what is sometimes found, three orders of tentacles placed symmetrically, the middle one of each group of three smaller ones being longer than its two neighbours. In *Ascidia virginea* also the tentacles are very numerous, nearly 100, and are of two sizes placed alternately.

Each tentacle is practically an ingrowth of the connective tissue of the body-wall, covered by the epithelial lining of the front of the alimentary canal. It has consequently a connective tissue core containing muscle fibres and nerves, and one or more blood lacunæ continuous with those of the body-wall. The delicate epithelium with which it is covered contains some simple sensory cells. These tentacles not only act mechanically in preventing large objects from entering, but are also sensitive like the lobes of the apertures, although only scattered sensory cells and no specially differentiated sense-organs are present.

Behind the tentacles lies the plain or papillated pre-branchial zone (Pl. II., fig. 7, *p.br.z.*) bounded behind by a pair of parallel and closely placed ciliated ridges with a groove between — the peripharyngeal bands (*p.p.b.*) —

which encircle the anterior end of the branchial sac. The anterior band forms a complete ciliated ring, but the posterior is interrupted in the ventral and dorsal median lines: its ends becoming continuous respectively with the marginal folds of the endostyle (ventrally), and with the front of the dorsal lamina, where before joining, they bound a narrow triangular cavity lined by ciliated epithelium, the epibranchial groove (see Pl. II., fig. 7, above *d.l.*). Behind the peribranchial bands the proper wall of the branchial sac commences.

Branchial Sac and Atrium.

The wall of the branchial sac is penetrated by a large number of channels, through which blood flows. Some of these run in one direction and some in another, so as to form complicated but perfectly definite networks, which differ in their arrangement in different kinds of Ascidians. Between these blood channels there are clefts (the secondary gill-slits or "stigmata") in the wall of the branchial sac, by means of which the water from the interior passes into the large external or peribranchial cavity—the atrium.

The transverse section (Pl. II., fig. 2) shows how this atrium surrounds the branchial sac on all sides except the ventral, where the wall of the branchial sac becomes continuous with the body-wall. The right and left halves of the atrium may be called the right and left peribranchial cavities (*p.br.*). They unite along the dorsal edge to form the cloaca, and there open to the exterior. The cavity of the branchial sac communicates with the surrounding atrium by means of the stigmata, as shown on the upper half (left side) of fig. 2 (Pl. II.). The section on the right side is shown passing along a transverse vessel between two of the rows of stigmata.

If an Ascidian expanded in sea-water, in a healthy condition, be closely watched, it will soon be noticed that there is a constant stream of water pouring in through the branchial aperture, and another flowing out from the atrial (as represented in fig. 1, Pl. II.); and if some fine and insoluble coloured powder be dropped into the water near the branchial aperture, it will rapidly be drawn in with the current, and after a short time some of the powder will make its appearance in the water ejected from the atrial aperture.

Hence, it is obvious that there is a current of sea-water flowing through the body of the Ascidian. This current has four distinct uses or functions: (1) it enables the animal to breathe, by bringing in fresh supplies of oxygen; (2) in the first part of its course it carries the microscopic food particles into the Ascidian's body; (3) in the last part of its course it carries out of the body various waste materials, which must be got rid of; and (4) it ejects the mature ova and spermatozoa from the body. The course which this water-current takes is:—in through the branchial aperture to the branchial sac, then through the clefts in the wall of that organ into the surrounding atrium, and lastly out through the atrial aperture to the exterior. The direction of this current may occasionally be temporarily reversed; and, when the muscular body-wall contracts, a sudden current may be ejected through both apertures simultaneously.

All the clefts or stigmata in the wall of the branchial sac (shown in fig. 3, Pl. II.) are bounded by cells which bear a number of cilia (Pl. IV., fig. 3) projecting across the cleft. These cilia, so long as the animal is alive, are in constant motion, lashing rapidly from the branchial sac towards the atrium, so as to drive the water in the cleft outwards; and it is this constant action of these very

14

minute cilia which causes the regular current of water to
flow in at the branchial aperture and out at the atrial.

The branchial sac is very large, much the largest organ
of the body (it may be 15 cm. in length), and extends
almost to the posterior end of the body, while the rest
of the alimentary canal lies upon its left side imbedded in
the body-wall. The food particles, consisting of very
minute plants and animals, are carried in through the
branchial aperture by the current of water, but most of
them do not pass out into the atrium, being caught by
the ciliary action of the peripharyngeal bands, and entangled
in the viscid substance which fills the groove between
them near the anterior end of the branchial sac. This
viscid substance, or mucus, is formed in a long canal-
shaped gland called the Endostyle or hypobranchial groove
(Pl. II., figs. 2 and 4, *end.*), which lies along the ventral
edge of the branchial sac, and terminates both anteriorly
and posteriorly in a short *cul-de-sac*.

THE ENDOSTYLE.

On the floor of the endostylar groove is found a tract
of cells with very long cilia (Pl. II., fig. 2). On each side are
at least two (in some Ascidians three) laterally placed
clumps of gland cells (Pl. IV., fig. 5), each clump separated
from its neighbours by an area of closely packed fusiform
cells with short cilia, amongst which are found some
sensory bipolar cells. The lips of the endostyle are formed
of ciliated cubical epithelium.

This organ, on account of its thick glandular walls,
has an opaque appearance, and seems, in side view, to
run like a conspicuous solid rod in the more transparent
walls of the branchial sac—hence its name endostyle. It
is, however, really a gland, and corresponds to the hypo-
pharyngeal groove of *Amphioxus* and the median part of

the thyroid gland of Vertebrata. It is interesting to
notice that the four (at least) longitudinal tracts of gland
cells are of remarkable constancy, being found not only in
all groups of Tunicata (often six tracts) including even the
pelagic, tailed, Appendicularians, but also in *Amphioxus*
and in the young thyroid gland of the Ammocoete.

The posterior *cul-de-sac* of the endostyle is quite short
in *Ascidia*, but in some other Tunicata it is longer and
becomes of great importance as an organ for the production of buds. Behind the *cul-de-sac* the marginal folds of
the endostyle unite to form a slight ridge, the posterior
fold, which runs round the posterior end of the branchial
sac to join the end of the dorsal lamina behind the
œsophageal opening. The endostyle, in addition to its
glandular function, shares in the sensory functions of the
peripharyngeal band, the tentacles, and the dorsal tubercle,
in all of which similar sensory cells and nerve endings have
been found.

DORSAL LAMINA.

The mucus formed by the glands in the lateral walls of
the endostyle is carried forwards by the lashing action of
the long cilia placed on the floor of the organ, and so
reaches the front of the branchial sac; here it changes
its direction of flow, and bends round to the right and left,
in the groove between the peripharyngeal bands, so as to
gain the dorsal edge of the sac, where it encounters a
projecting membranous fold, the dorsal lamina or epibranchial ridge, along which it is carried backwards—still
by ciliary action—to the opening of the œsophagus, the
next region of the alimentary canal after the pharynx.
The food particles become entangled in this train of mucus
chiefly in its course round the right and left sides of the
anterior end of the sac, and from this point are carried

back along the right side of the dorsal lamina to the oesophagus.

The dorsal lamina is at its widest round the oesophageal aperture (Pl. II., fig. 4, *oes.*). It is more or less ridged transversely, especially on its convex left side, and may have marginal tags or teeth (Pl. IV., fig. 4), which in some Ascidians become long processes, the languets. In the living animal, the lamina has its free edge curved to the right hand side in such a manner as to constitute a fairly perfect tube along which the train of food passes.

Wall of Branchial Sac.

Fig. 3, Pl. II. shows a small part of the wall of the branchial sac of *Ascidia* in which it may be seen that the bars containing the blood channels are arranged in three regular series :—

(1) The "transverse vessels" (*tr.*) which run horizontally round the wall and open at their dorsal and ventral ends into large median longitudinally running tubes, the dorsal blood sinus behind the dorsal lamina and the ventral blood sinus below the endostyle (see also Pl. III., fig. 10).

(2) The fine longitudinal or "interstigmatic vessels" (*l.r.*) which run vertically between adjacent transverse vessels and open into them, and which therefore bound the stigmata.

(3) The "internal longitudinal bars" (*i.l.*) which run vertically in a plane internal to that of the transverse and fine longitudinal vessels. These bars (see Pl. II., fig. 3, *B.*) communicate with the transverse vessels by short side branches (*c.d.*) where they cross, and at these points are prolonged into the cavity of the sac in the form of hollow papillæ (*p.*).

In some Ascidians (*e.g. Corella parallelogramma* and most of the Molgulidae) the interstigmatic vessels are curved

17

so that the stigmata form more or less complete spirals. In some species of *Ascidia*, and other Ascidians, the inter-stigmatic vessels are inserted into the transverse ones in undulating in place of straight lines, the result being that the stigmatic part of the wall of the branchial sac seems to be folded or thrown into microscopic crests and troughs. This device for increasing the surface is known as "minute plication," and is seen well in *Ascidia mentula* (Pl. IV., figs. 1 and 2). In some cases, again (Cynthiidæ), the whole wall of the sac is pushed inwards at intervals to form large folds visible to the eye.

The intersections of the internal longitudinal bars with the transverse vessels divide up the inner surface of the branchial sac into rectangular areas called "meshes" (Pl. IV., fig. 2). One such mesh, containing eight stigmata in a row, is seen in fig. 3, Pl. II. The internal longitudinal bars bear hollow papillæ at the angles of the meshes, and occasionally in intermediate positions. There are frequently horizontal membranes (fig. 3, B. *h.m.*) attached to the transverse vessels between the papillæ, and the transverse vessels may be of two or more sizes arranged symmetrically. There are many "connectives" running from the outer wall of the branchial sac to the body-wall, and allowing the blood in the transverse vessels to communicate with that in the sinuses outside (Pl. II., fig. 2, *con.*).

In an adult medium-sized *Ascidia mentula* there are about 150 transverse vessels on each side of the branchial sac, and at least 80 internal longitudinal bars, making by their intersection 12,000 meshes. The average number of stigmata in a mesh is, in this species, six to eight. So there may be as many as 96,000 stigmata present on each side of the sac, nearly 200,000 in all. Probably these numbers are greatly exceeded in large specimens, such as that shown on Pl. I. These stigmata are to be regarded as secondary openings due to the breaking up or subdivision of the primary Chordate gill-clefts.

In addition to the stigmata there are generally one or two pairs of much larger (up to 15 mm. in length), narrow, elongated slits placed near the posterior end of the sac and close to the dorsal lamina, so as to be underneath the atrial aperture, by which water can escape into the cloacal part of the peribranchial cavity. These pharyngo-cloacal slits have well-marked edges which bear much finer cilia than those of the stigmata.

The lining of the branchial sac is the pharyngeal epithelium (endoderm) while the outer surface is covered by the lining membrane of the peribranchial cavity in free communication with the outer surface of the body through the atrial aperture (see Pl. II., fig. 2). Both of these epithelial surfaces are formed of squamous cells. Round the sides of the stigmata (Pl. IV., fig. 3) the cells on the longitudinal vessels become more nearly cubical in shape and bear the cilia, while at the ends of the stigmata, near the transverse vessels, the cells approach a columnar form (Pl. IV., fig. 3). The epithelium along the internal edge of the longitudinal bars, and on the apices of the papillæ, is also cubical or almost columnar in form.

Between the outer and the inner epithelium the wall of the branchial sac is formed of connective tissue in which the blood lacunæ, known as "vessels," are excavated. These vessels are very regular in size and arrangement, and are so large that comparatively little connective tissue is left, and so the blood is in close proximity with the epithelial surface. In some branchial sacs a few non-striped muscle fibres are found running longitudinally in the connective tissue around the chief vessels (see Pl. IV., fig. 1).

ALIMENTARY CANAL.

The œsophagus is a short, narrow, curved tube which leads ventrally to the stomach—a large, thick-walled organ, lying on the left-hand side of the branchial sac, imbedded in the body-wall and projecting into the atrium, (as shown in fig. 4, Pl. II.). From the other (ventral) end of the fusiform smooth-walled stomach arises the intestine, a long serpentine tube which ends by opening into the dorsal or cloacal part of the atrium, from which the undigested portions of the food are carried to the exterior through the atrial aperture, by the water current.

The intestine is curved so as to form two loops (Pl. II., fig. 4), a first between the stomach and intestine, open posteriorly, and in which the ovary lies; and a second between the intestine and rectum, open anteriorly, and in which the renal vesicles lie. The external convex edge of the intestine is thickened internally to form the "typhlosole," a large pad which runs along its entire length (Pl. II., figs. 2 and 4, *ty.*), reducing the lumen to a crescentic slit.

The walls of the stomach are glandular; and, in addition, a system of delicate hyaline microscopic branched tubules with dilated ends (the "refringent organ"), which ramifies over the outer wall of the intestine and communicates with the cavity of the stomach at the pyloric end by means of a duct, is possibly a digestive gland. There is in *Ascidia* no separate large gland to which the name "liver" can be applied as in some other Tunicata (*e.g., Molgula*). Over these viscera, on the left side of the body, the body-wall is thin and gelatinous, and has usually no muscle fibres visible.

The wall of the alimentary canal, throughout its length, consists of an epithelial lining (endoderm), a thick layer of highly vascular connective tissue, and, over that, the flattened epithelium lining the peribranchial cavity. The only notable regional differences lie in the epithelium. The connective tissue is much the same throughout, and is continuous with that of the body-wall in which the digestive viscera are imbedded. The blood lacunæ are very numerous and communicate on the one hand with the cardio-visceral, and on the other hand with the branchio-visceral main blood vessels.

The connective tissue is also penetrated by (*a*) the cæca of the gonads, especially the spermatic tubules, and (*b*) the delicate branched clear tubules of the enigmatical refringent organ or pyloric gland (Pl. IV., fig. 7). The tubules of this organ are lined by low cubical cells, usually non-ciliated, containing no concretions or granules, and having no great resemblance to gland cells. They have been called " chylific " and absorptive, but the function is still undetermined. There are few, if any, muscle fibres in the connective tissue of the alimentary canal until the rectum is reached.

The lining epithelium is for the most part ciliated, especially in the œsophagus and intestine. The œsophagus has slight longitudinal ridges and grooves, one of which seems to continue the canal of the dorsal lamina onwards to the stomach, while another is in relation with the posterior fold coming from the lower end of the endostyle. There are a few glandular (mucous) cells scattered amongst the ciliated columnar cells of the œsophageal wall.

The stomach has projecting folds in its interior which unite at its pyloric end to form the intestinal typhlosole, and, in addition to some ciliated and mucous cells, it has large masses of highly coloured markedly glandular cells packed with yellow granules. The intestine, again, has ciliated and gland cells, and in the rectum the gland cells become fewer and die out. The termination of the rectum, close to the atrial aperture, is very thin-walled, with a slight thickened edge at the anus, but no sphincter.

HEART AND CIRCULATION.

The soluble product of the food which has been digested passes through the wall of the alimentary canal, and enters the numerous small blood spaces in the connective tissue on both sides of the stomach and intestine. These lead, by the cardio-visceral vessels, to the dorsal end of the heart (see Pl. III., fig. 10), which is merely a delicate tube, irregularly swollen in the middle, placed behind the stomach, and projecting into a space, the pericardium (Pl. IV., fig. 9, *p.c.*), which is a part of the original cœlom.

The wall of the heart is continuous along one edge (that next the stomach) with that of the pericardium, and the heart is to be regarded as a tubular invagination of the pericardial wall (see Pl. IV., fig. 10), shutting in a portion of the external space (the blastocœle of the embryo) and having open ends which communicate with the large blood sinuses leading to the branchial sac, to the viscera, and to the body-wall and test. The cavity of the heart is not divided and has no valves. Its wall is formed of a single layer of epithelio-muscular cells, the inner (muscular) ends of which are cross-striated fibres running round the heart —the only striated muscle found in the body of the Ascidian. The larger channels through which the blood flows are lined by a delicate endothelium, the smaller are merely spaces in the connective tissue. All the blood spaces and lacunæ are probably derived, like the cavity of the heart, from the blastocœle of the embryo, and are not (like the pericardium) a derivative of the cœlom. The wall of the pericardium is simple squamous epithelium.

From the ventral end of the heart the blood is conveyed by the branchio-cardiac vessel and the great ventral vessel lying underneath the endostyle (Pl. III., fig. 10, *br. ao.*) to the walls of the branchial sac, where, in passing along the transverse and interstigmatic vessels, it is purified, and receives a supply of oxygen from the water passing through the stigmata. It is then conveyed, by the great dorsal vessel and the branchio-visceral vessel and its branches, to different parts of the viscera and body-wall, so that all the organs may receive food and oxygen, and have their waste materials carried away. Some of these branchio-visceral vessels from the branchial sac lead to the walls of the stomach and intestine (see Pl. III., fig. 10), and thus bring us back to the point from which we started.

The great dorsal and ventral vessels of the branchial sac are connected (Pl. III., fig. 10), not only by the transverse vessels, which run like hoops round the walls, but also at their anterior extremities, by a circular vessel which surrounds the front of the branchial sac, underneath the peripharyngeal bands. A short branch runs from near the front of the dorsal vessel to the sinuses which surround the nerve ganglion.

From each end of the heart "vessels" are also given off to supply the body-wall and test. Moreover, "connectives" run from the transverse vessels of the branchial sac directly outwards, on each side, to the body-wall and viscera. On the left side there are three especially large blood tubes amongst the connectives, which cross to the alimentary canal and ovary, and branch through these viscera.

But the course of this circulation of the blood is not always the same; sometimes it is exactly reversed, the blood flowing from the branchial sac to the heart, and from that organ to the viscera, and then back to the branchial sac again. This curious state of affairs is caused by the remarkable manner in which the Ascidian heart beats. If a young individual of a small transparent species, such as *Ascidia virginea*, be placed alive, left-side uppermost, in a watch-glass of sea-water, and examined with a low power of the microscope, the heart will be readily seen near the posterior end of the transparent body. It will then be noticed that the "beating" looks like successive waves of blood, which are pressed through the tubular heart from one end to the other by the contractions of the muscle fibres. After watching the waves passing, let us suppose, from the dorsal end of the heart to the ventral, for about a minute and-a-half or two minutes, it will be seen that they gradually become slower and then stop altogether. But now, after several seconds, a faint wave will start from the *ventral* end of the heart and pass over it to the dorsal; and this will be followed by larger ones for perhaps a minute or two, and then again a pause will occur and the direction change. So that we may say, the heart of the Ascidian beats first in one direction and then in the other; and the reversal of the blood current takes place every minute or two. There are generally rather more beats in the dorso-ventral than in the opposite direction, but there is considerable irregularity. The numbers are usually between 60 and 80.

The cause of this remarkable reversal may possibly be that the heart being on the ventral vessel, which is wider than the corresponding dorsal trunk, it pumps the blood into either the lacunæ of the branchial sac or those of the viscera in greater volume than can possibly get out through the smaller branchio-visceral vessel in the same time, the result being that the lacunæ in question will soon become engorged, and by back pressure cause the stoppage, and then reversal of the beat. The absence of any valves in the heart to regulate the direction of flow obviously facilitates this alternation of the current.

When the heart is contracting ventro-dorsally it receives oxygenated blood from the branchial sac by the branchio-cardiac vessel (now a vein), and propels it by the cardio-visceral trunk (now an artery) to both sides of the viscera and body-wall. This blood, after circulating through the system, is collected as impure blood by the branchio-visceral vessel and conveyed to the dorsal sinus of the branchial sac to be re-oxygenated. The heart is then a systemic heart and contains pure blood. But after the reversal, when the heart contracts dorso-ventrally the veins and arteries exchange functions, the oxygenated blood passes from the branchial sac to the viscera, the heart receives impure blood from the system and propels it to the ventral edge of the branchial sac, and so what was a minute before a " systemic," is now a " respiratory" heart. This is a phenomenon without parallel in the animal kingdom.

The blood of Ascidians is in the main transparent, but contains usually certain pigmented corpuscles in addition to many ordinary leucocytes or colourless amœboid nucleated cells (Pl. IV., fig. 6). The pigment in the coloured cells may be red, yellow, brown, or in some cases blue or opaque white, and these are the result of deposition of pigment granules in the older leucocytes. In *Ascidia mentula* a large number of blood corpuscles are usually brown. The unaltered leucocytes may be actively amœboid, and can proliferate. As we have seen, the blood may reach the branchial sac either from the dorsal or from the ventral median sinus, according to the direction in which the heart is beating at the moment ; and it is a most interesting and beautiful sight to watch the alternating circulation of the variously coloured corpuscles through the transparent vessels, and the lashing of the cilia along the edges of the

neighbouring stigmata, as shown in a small Ascidian under
the microscope.

NERVOUS SYSTEM AND SENSE ORGANS.

The nervous system of the Ascidian consists of a single
elongated Ganglion or " brain," placed near the front end
of the body, in the dorsal median line, between the
branchial and atrial apertures (Pl. II., fig. 4, *n.g.*) It
gives off several large nerves at each end, which break up
into the fibres that go to the different parts of the body,
and especially to the lobes and the muscles surrounding
the branchial and atrial apertures (Pl. II., fig. 6). In
Ascidia mentula there are four chief nerves from the
anterior end of the ganglion, one to the surface of the
body and the other three to the branchial siphon; while
three nerves run from the posterior end to the atrial
aperture and wall of cloaca. There are also one or two
smaller lateral nerves that leave the sides of the ganglion.

This ganglion is the degenerate remains of the anterior
part of the cerebro-spinal nervous system of the tailed
larval Ascidian. The posterior or spinal part has almost
entirely disappeared in most adult Tunicata; but its
remains may be traced in a tract of degenerate nerve
tissue (the dorsal nerve cord) which runs posteriorly from
the ganglion (Pl. II., fig. 6, *d.n.*) above the base of the
dorsal lamina towards the viscera. The ganglion has
small rounded nerve cells on its surface, while the centre
is a mass of interlacing nerve fibres. Small cells are
also found scattered along the course of the dorsal nerve
cord. The nerve cells in *Ascidia mentula* are mostly
pyriform or triangular in form, and are bi-polar or multi-
polar, and finely granular.

The Ascidian has little papillæ containing sensory cells
in its ectoderm, especially round the apertures, and has

isolated sense-cells and nerve-endings in various parts of its internal epithelium; but is very badly provided with more definite "sense organs." It has no true eyes—for the little brightly-coloured dots or "ocelli" placed along the margins of the apertures (Pl. I.), and formed of a group of modified ectoderm cells, supplied by a nerve and imbedded in a mass of red and yellow pigment (Pl. IV., fig. 8), can scarcely be called such; and it certainly has no ears or otocysts. The tentacles are not very efficient tactile organs, and the thin expanded margins of the branchial and atrial siphons are apparently the most sensitive parts of the body.

But there is a curiously-curled projection, the Dorsal Tubercle (Pl. II., fig. 7, d.t.), placed at the front of the dorsal lamina, in the prebranchial zone, near the entrance to the branchial sac, which may possibly be an organ for testing, by smell or taste, the quality of the water drawn in through the branchial aperture. This organ has a narrow slit which leads by means of a ciliated funnel into a delicate non-ciliated tube, and this can be traced back for a distance of several centimetres to a glandular mass, the neural gland, formed of tubules lined by small cubical cells, lying imbedded in the connective tissue immediately underneath the brain (see n. gl., fig. 6, Pl. II.).

Hence, it has been suggested that the supposed olfactory organ is merely the complicated opening of the duct from the neural gland, and that this gland probably corresponds to the hypophysis cerebri or pituitary body, which is found in all Vertebrata, from fishes up to man, attached to the infundibulum on the lower surface of the brain. It is probable that both views are partly right, and that therefore the duct of the pituitary gland in the Ascidian, opens into a sense organ placed on the roof of the mouth.

Sensory cells have been found amongst the ciliated

epithelium of the dorsal tubercle, as well as in the tentacles, the peripharygneal bands, the endostyle, and neighbouring parts.

As to the function of the neural gland—apart from the dorsal tubercle—it is still somewhat mysterious. It may be merely to secrete viscid matter which is poured like that from the endostyle into the peripharygneal groove, or it may possibly be that the function is renal—for the removal of nitrogenous waste matters in the neighbourhood of the nervous system.

RENAL ORGAN.

A mass of large clear-walled vesicles, which occupies the rectal loop and the adjacent walls of the intestine, and may extend over the whole left side, is undoubtedly a renal organ without a duct. Each vesicle (Pl. IV., fig. 11) is apparently a little closed sack formed of modified mesoblast cells which eliminate nitrogenous waste matters from the blood in the neighbouring lacunæ and deposit them in the cavity, where they form one or more constantly increasing concentrically laminated concretions of a yellowish or brown colour, sometimes coated with a chalky deposit. These concretions (Pl. IV., fig. 14) contain uric acid, and in a large Ascidian are very numerous and of considerable size. The nitrogenous waste products are thus deposited and stored up throughout life in the renal vesicles in place of being excreted from the body.

The cells forming the walls of the renal vesicles have a curiously wavy outline (Pl. IV., figs. 11, 12, and 13), which gives them a characteristic appearance. The contents of these cells seem to differ considerably (Pl. IV., fig. 13) in different cases, probably as a result of their functional activity.

REPRODUCTIVE ORGANS.

Ascidians are not divided into two sexes, and consequently each individual is hermaphrodite, or possesses both male and female reproductive organs (gonads), although cross - fertilization is probably the rule as a protogynous condition is very general—the female organs maturing before the male. In *Ascidia mentula*, however, self-fertilization does sometimes take place. The gonads in *Ascidia* lie close together on the left side of the body, alongside the stomach and intestine (Pl. II., fig. 4, *gon.*), and are provided with delicate ducts (*gd.*), which open like the intestine into the atrium; so that the mature ova and spermatozoa are carried out of the Ascidian's body by the current of water flowing from the atrial aperture. In many Ascidians fertilisation and development take place in the atrium, a part of which may be set aside as an incubatory pouch. In some Ascidians (certain Molgulidæ and Cynthiidæ) reproductive organs are present on both sides of the body, and in others (*Polycarpa*) there are many complete sets of both male and female systems attached to the inner surface of the body-wall, on both sides, and projecting into the peribranchial cavity.

The ovary in *Ascidia mentula* is a slightly ramified gland which occupies the greater part of the intestinal loop (Pl. II., fig. 4, *gon.*). It contains a cavity which, along with the cavities of the testis, is derived from an embryonic mesodermal space which has been compared with a cœlom but may be merely blastocœlic, and the ova are formed on its walls and fall when mature into this cavity. The oviduct is directly continuous with the cavity of the ovary, and leads forward alongside the rectum, and external to the vas deferens to open into the atrium.

The testis is composed of a great number of delicate branched white tubules, which ramify over the much coarser tubules of the ovary and the adjacent parts of the intestinal wall. These spermatic tubules terminate in ovate swellings, usually grouped in bunches (Pl. IV., fig. 15). Near the commencement of the rectum the larger tubules unite to form the vas deferens, a tube of considerable size which runs forward alongside the rectum, and, like the oviduct, terminates by opening into the peri-branchial cavity close to the anus. The lumen of the tubules of the testis, like the cavity of the ovary, is a meso-blastic space in the embryo, and the spermatozoa are formed from the cells lining the wall, and are set free into the cavity.

The mature ovum is of small size (about 0·12 mm. in diam.), colourless, and with little or no food yolk in the case of *Ascidia*. It is only some of the germinal cells in the ovary that are destined to become ova. Of the rest, some form a protecting layer, the follicle, around the young ova. Certain of these primary follicle cells migrate inwards and give rise, by proliferation, to a layer of cells in the superficial part of the ovum (Pl. V., fig. 1, *t.c.*). These are the so-called "testa-cells" or kalymmocytes. These later on produce a thin gelatinous layer over the surface of the ovum and between it and the follicle, which looks like the beginning of the test—hence the name given to the cells, which, although so different in origin from the mesodermal test cells of the adult, probably to some extent give rise to "bladder cells" in the test. The rest of the follicular "testa-cells" eventually disappear.

The follicle cells proper produce two layers, the outer of which remains in the wall of the ovary when the ovum is set free; while the inner layer adheres to the surface (Pl. V., fig. 1, *foll.*), and its cells become large and much vacuo-lated, some of them growing out to form long papillæ, which help to sustain the floating egg in the sea-water.

Altogether there may be as many as seven distinct layers around the mature egg, but they are all produced by the differentiation or activity of the follicle cells.

Polar bodies are formed from the maturing ovum in the usual manner, and effect the usual reduction in the number of chromosomes in the nucleus. In the common *Styelopsis grossularia* there are two chromosomes left in the ovum, while in some species of *Ascidia* there are eight.

The spermatozoa of *Ascidia* are of typical form (see Pl. V., fig. 1, s.).

EMBRYOLOGY AND LIFE-HISTORY.

The egg (Pl. V., fig. 1) after being fertilised (probably in most cases by a spermatozoon carried by the current of water from *another* Ascidian somewhere in the neighbourhood), proceeds to segment or divide into a number of small pieces or young cells, thus becoming an embryo Ascidian (Pl. V., fig. 2, &c.). The cells of the embryonic body gradually come to be arranged (in a manner the details of which will be described below) so as to form — (1) a skin or layer of cells—the ectoderm of the adult—covering the outside; (2) a tubular nervous system running along the middle of the dorsal surface, underneath the ectoderm; (3) a short wide tube, placed ventrally, which is the beginning of the branchial sac and the remainder of the alimentary canal; and (4) a cellular rod—the notochord—which lies in the posterior part of the body, between the dorsal nerve tube and the ventral alimentary canal, and is a rudimentary or very simple back-bone, similar to that found in the embryos of vertebrate animals (fig. 6, Pl. V.). In fact, the embryo Ascidian at this stage is comparable with an embryo fish or frog, and is found to have the same chief organs or parts similarly arranged; and, moreover,

these parts have been formed in essentially the same
manner in both cases; so that if similarity in structure
and development indicate relationship, it is evident that
the young Ascidian is related to the fish and the frog and
other Vertebrata, and is to be regarded as one of the
Chordata (animals which at some time of their life have a
notochord).

Turning now to the details of the development,[*] the
segmentation is complete, and bilateral, and nearly equal,
and results in the formation of a spherical blastula with
a small segmentation cavity (Pl. V., fig. 2).

The blastula then grows larger and begins to differen-
tiate. There are slightly smaller cells which divide more
rapidly at one end of this embryo, the future ectoderm,
and slightly larger and more granular cells at the other,
which become chiefly endoderm (hypoblast). Invagination
of the larger cells then takes place (fig. 3), resulting in the
formation of a gastrula with an archenteron. The hypoblast cells lining the archenteron become columnar (hy.).
The curving and more rapid growth at the anterior end of
the embryo narrow the primitively wide open blastopore,
and carry it to the posterior end of the future dorsal
surface (Pl. V., fig. 4, b.p.). The directions of the body
are now clear. The embryo is elongated antero-posteriorly,
the dorsal surface is flattened and the blastopore indicates
its posterior end. Around the blastopore certain of the
ectoderm cells form a medullary plate along which a
groove (the medullary groove) runs forwards, bounded
at the sides by laminæ dorsales which meet behind the
blastopore. Underneath the posterior part of the medullary
groove certain of the hypoblast cells from the dorsal wall
of the archenteron in the median line form a band

[*] The early stages of *Ciona*, of which Castle has given a very complete
account, differ in some points from those of *Ascidia* described here.

extending forward (fig. 4, *ch.*). This band separates off from the hypoblast, which closes in beneath it, and thus the notochord is formed (fig. 5, *ch.*). The same cells further laterally and posteriorly become mesoblast, and separate off as lateral plates which show no trace of metameric segmentation (fig. 7a, *m.b.*). The remainder of the archenteron will become the branchial sac, and by further growth bud off the rest of the alimentary canal.

The medullary groove now becomes converted into the closed neural canal by the growing up and arching inwards (fig. 7a, *n.c.*) of the laminæ dorsales, which unite with one another from behind forwards in such a way that the blastopore now opens from the enteron into the floor of the neural canal, forming the neurenteric passage (fig. 5, *n.c.c.*). For a time the anterior end of the neural canal remains open as a neuropore. The posterior end of the body is now elongating to form a tail, and the embryo is rapidly acquiring the tadpole shape (fig. 6) characteristic of the free larva.

The tail grows rapidly, curves round the body, and also undergoes torsion so that its dorsal surface comes to lie on the left side. It contains ectoderm cells on its surface, notochordal cells (in single file) up the centre (see fig. 7, *n.ch.*), a neural canal dorsally, and a row of endoderm cells (*hy.*) representing the enteron ventrally to the notochord. Later on the mesoblast also is prolonged into the tail where it forms a band of striated muscle-cells at each side of the notochord. When the ectoderm cells begin to secrete the cuticular test it forms two delicate transparent longitudinal (dorsal and ventral) fins in the tail (fig. 7), and especially at its extremity where radial thickenings form striæ resembling fin rays. The ectoderm on the anterior end of the body grows out into three adhering papillæ (figs. 8 and 9).

The neural canal now differentiates into a tubular dorsal nervous system. The anterior end dilates to form the thin-walled cerebral vesicle (figs. 8 and 9, *n.v.*), containing later the intra-cerebral, dorsal, pigmented eye (*oc.*) and the ventral otolith (*au.*) of the larva. The next part of the canal thickens to form the trunk ganglion, and behind that is the more slender spinal cord (*n.c.*), which runs to the extremity of the tail. A ciliated diverticulum of the anterior end of the enteric cavity (future pharynx), which enters into close relations with the front of the cerebral vesicle, and later opens into the ectodermic invagination which forms the mouth at that spot, is evidently the rudiment of the neural or hypophysial gland and canal.

The future branchial sac (fig. 9, *mes.*), with a ventral median thickening which will be the endostyle, is by this time clearly distinguishable, by its large size, from the much narrower posterior part of the enteron which grows out to become the œsophagus, stomach, and intestine.

The notochord does not extend forward into the branchial region, but is confined to the posterior or caudal part of the embryo. It now shows lenticular pieces of a gelatinous intercellular substance secreted by the cells and lying between them (fig. 8). The mouth forms as a stomodæum or ectodermal invagination antero-dorsally, in the region where the neuropore had closed up; and, about the same time, two lateral ectodermal involutions appear (fig. 9, *at.*) which become the atrial or peribranchial pouches, at first distinct, afterwards united in the mid-dorsal line to form the adult cloaca and atrial aperture.

In-growths from the atrial pouches and out-growths from the wall of the pharynx coalesce to form the proto-stigmata (primary gill slits) by which the cavity of the branchial sac is first placed in communication with the exterior through the atrial apertures. Opinions differ as to whether only one or a few pairs of true gill clefts are represented in the young Ascidian, and the actual details of their formation and sub-division differ greatly in different forms. To what precise extent the walls of the atrial or peribranchial cavities are formed of ectoderm or endoderm, is also still doubtful.

The embryo is hatched, about two or three days after fertilisation, as a larva or Ascidian tadpole (fig. 9) which leads a free-swimming existence for a short time during which it develops its nervous system and cerebral sense organs, and the powerful mesoblastic muscle bands lying at the sides of the notochord (now a cylindrical rod of gelatinous nature surrounded by the remains of the original cells) in the tail, and forming the locomotory apparatus. Figure 9 shows this stage, the highest in its chordate organisation, when the larva swims actively through the sea by vibrating its long tail provided with dorsal and ventral fins.

In addition to the structures already mentioned, the mesoderm has formed the beginning of the muscular body-wall and the connective tissue around the organs, and has given rise to the blood, the endostyle has developed as a thick-walled groove along the ventral edge of the pharynx, which now becomes the branchial sac, and the pericardial sac and its invagination the heart have formed in the mesoblast between the endostyle and stomach. The unpaired optic organ in the cerebral vesicle, when fully formed, has a retina, pigment layer, lens and cornea; while the ventral median sense-organ is a large spherical, partially pigmented otolith, supported by delicate hair-like processes on the summit of a hollow "crista acustica" (fig. 9). Both the otolith and the retina and lens of the eye are formed originally by the differentiation of a group of cells in the epithelium lining the cerebral vesicle—they are myelonic sense-organs.

After a few hours, or at most a day or so, the larva attaches itself by one or more of the three anterior ecto-dermal glandular papillæ (one dorsal and two lateral) to some foreign body, and commences the retrogressive metamorphosis which leads to the adult state. The adhering papillæ having performed their function begin to atrophy, and their place is taken by the rapidly increasing test. The tail, which at first vibrates rapidly, is partly withdrawn from the test and absorbed and partly cast off in shreds (figs. 10 to 12). The notochord, nerve tube, muscles, &c., are withdrawn into the body, where they break down and are absorbed by phagocytes, or dissolved in the fluid of the body-cavity. The posterior part of the nerve tube and its anterior vesicle with the large sense organs disappear, and the middle part undergoes prolifera-tion dorsally to form the relatively small ganglion of the adult, underneath which the neural tube gives rise to the hypophysial gland. While the locomotory, nervous, and sensory organs are thus disappearing or being reduced, the alimentary canal and reproductive viscera are growing larger. The branchial sac enlarges, its walls become penetrated by blood channels and grow out to form bars and papillæ, and the number of openings greatly increases by the primary gill slits becoming broken up into the transverse rows of stigmata.

The stomach and intestine, which developed as an out-growth from the back of the branchial sac at the dorsal edge, become longer and curve so that the end of the intestine acquires an opening into at first the left-hand side, and eventually the cloacal or median part of the atrial cavity. The adhering papillæ have now disappeared, and are replaced functionally by a growth of the test over neighbouring objects; and, at the same time, the region of the body between the point of fixation and the mouth (branchial aperture) increases rapidly in extent so as to cause the body of the Ascidian to rotate through about 180°, and so carry the branchial siphon to the opposite end from the area of attachment (see figs. 10, 11, 12, and 13 on Pl. V.).

Finally, the gonads and their ducts form in the mesoderm between stomach and intestine, and so bring us to the sedentary degenerate fixed adult Ascidian with little or no trace of the Chordate characteristics so marked in the earlier larval stage (compare figs. 13 and 9). The free-swimming tailed larva shows the Ascidian at the highest level of its organisation, and is the stage that indicates the genetic relationship of the Tunicata with the Vertebrata. In some Ascidians with more food-yolk in the egg, or in which the development takes place within the body of the parent, the life-history as given above is more or less modified and abbreviated, and in some few forms the tailed larval stage is missing.

The remarkable life-history of the typical Ascidian, of which the outlines are given above, is of importance from two points of view:—

1st. It is an excellent example of degeneration. The free-swimming larva is a more highly developed animal than the adult Ascidian. The larva is, as we have seen, comparable with a larval fish or a young tadpole, and so is a chordate animal showing evident relationship to the Vertebrata; while the adult is in its structure non-chordate, and may be regarded as being *on a level* with some of the worms, or with the lower Mollusca, in its organisation—although of an entirely different type.

2nd. It shows us the true position of the Ascidians (Tunicata) in the animal series. If we knew only the adult forms we might regard them as being an aberrant group of the Vermes, or possibly as occupying a position between worms and the lower Mollusca, or we might place them as an independent group; but we should certainly have to class them as Invertebrate animals. But when we know the whole life-history, and consider it in the light of "recapitulation" and "evolutionary" views, we recognise that the Ascidians are evidently related to the Vertebrata, and were at one time free-swimming Chordata occupying a position somewhere below the lowest Fishes.

N.B.—In an account of this nature, in which I have obviously made the fullest use of the published works of my predecessors, I have not considered it necessary to burden the text with frequent references to original memoirs. While accepting, then, full responsibility for my statements—nearly all of which I have taken occasion to verify by personal observation—I do not, of course, claim any originality in regard to them. Several excellent bibliographies of the Tunicata have already been published : another seems superfluous.

APPENDIX.

It may be useful to add here a brief statement of the classification and characters of the TUNICATA, in order to indicate the position of *Ascidia* as a type of the group, and its relations to the other British Ascidians.*

TUNICATA.

The Tunicata (or Urochorda) are hermaphrodite marine chordate animals, which show in their development the essential vertebrate characters, but in which the notochord is restricted to the posterior part of the body, and is in most cases present only during the free-swimming larval stage. The adult animals are usually sessile and degenerate, and may be either solitary or colonies, fixed or free. The nervous system is in the larva of the elongated, tubular, dorsal vertebrate type, but in most cases degenerates in the adult to form a small ganglion placed above the pharynx. The body is completely covered with a thick cuticular test ("tunic"), which contains a substance similar to cellulose. The alimentary canal has a greatly enlarged respiratory pharynx (the branchial sac), which is perforated by two, or many, more or less modified gill slits, opening into a peribranchial or atrial cavity, which communicates with the exterior by a single dorsal exhalent aperture, rarely two ventral apertures. The ventral heart is simple and tubular, and periodically reverses the direction of the blood current.

* For a more detailed classification, with definitions of all the groups and analytical keys to the species, see Herdman's Revised Classification of the Tunicata. Journ. Linn. Soc. Zool., vol. XXIII, p. 558, 1891.

The leading vertebrate characteristics of the Tunicata are the notochord, the dorsal nervous system, the ventral heart, and the respiratory pharynx with gill clefts, but these all disappear or undergo modification to such an extent, during the metamorphosis, that the degenerate adults would not, in most cases, be recognised as belonging to the chordata were it not for our knowledge of the life-history.

The class Tunicata may be divided into three orders :—

Order I. LARVACEA.

This comprises the free-swimming, permanently-tailed, larva like, mostly minute Appendicularians. A relatively large test or " House " is formed with great rapidity as a secretion from the surface of a special part of the ectoderm ; it is, however, merely a temporary structure, which may be cast off and afterwards replaced by another " House." The branchial sac is simply an enlarged pharynx, with two ventral ciliated openings (stigmata) leading to the exterior. These open independently on the ventral surface, and there is no separate peribranchial cavity. The tail is a large locomotor appendage, in which there is a skeletal axis, the urochord, comparable with the notochord of Vertebrata. The nervous system consists of a large anterior and dorsally-placed ganglion, and a long nerve cord with smaller ganglia stretching backwards from it over the alimentary canal to reach the tail, along which it runs on the left side of the urochord. The alimentary canal lies behind the branchial sac, and the anus opens ventrally on the surface of the body in front of the stigmata (or atriopores). The gonads are placed at the posterior end of the body. Gemmation does not take place, and alternation of generations and metamorphosis do not occur in the life-history.

This group contains a single family, the Appendiculariidæ, all minute (about 5 mm. long), tailed, free-swimming forms which have undergone comparatively little degeneration, and, consequently, correspond more nearly to the tailed larval condition than to the adult forms of the other groups of Tunicata. There are nearly a dozen genera known, of which at least two, probably more, inhabit British seas. In the genus *Oikopleura*, to which our commonest Appendicularians belong, the body is short and ovoid, and no anterior fold or "hood" is present. The tail is three or four times the length of the body, and four to six times as long as it is broad. In *Fritillaria*, on the other hand, the body is elongated, and somewhat constricted in the middle where the tail is attached. A fold of integument on the front of the body forms a "hood." The tail is short and wide, not twice as long as the body.

The British species of Larvacea are still insufficiently known.

Order II. THALIACEA.

Free-swimming pelagic forms of moderate size, which may be either Simple or Compound, and in which the adult is never provided with a tail or notochord. Consequently the whole body here corresponds to the trunk only of the Appendicularian, without the tail. The test is permanent, and may be either well-developed or very slight. In all cases it is clear and transparent. The musculature of the body-wall is in the form of more or less complete circular bands, by the contraction of which water is ejected from the body, and so locomotion is effected. The branchial sac has either two large or many small apertures (stigmata), leading to a single peribranchial cavity, into which the anus also opens. Alternation of generations occurs in the life-history, and may be complicated by polymorphism. The order Thaliacea comprises two groups, CYCLOMYARIA and HEMIMYARIA.

Sub-order I. CYCLOMYARIA.

Free-swimming pelagic forms, which exhibit alternation of generations in their life-history, but never form permanent colonies. The body is cask-shaped, with the branchial and atrial apertures at the opposite ends. The test is moderately well developed. The musculature is mostly in the form of complete circular bands surrounding the body. The branchial sac is fairly large, occupying the anterior half or more of the body. Stigmata are usually present in its posterior part only. The peribranchial cavity is mainly posterior to the branchial sac. The alimentary canal is placed ventrally close to the posterior end of the branchial sac. Hermaphrodite reproductive organs are placed ventrally near the intestine.

This group contains only one British genus, *Doliolum*, and even that is probably only an occasional visitant. It has a cask-shaped body, from 1 to 2 cm. in length, with lobed terminal branchial and atrial apertures. The body-wall contains eight or nine circular muscle bands, by the contraction of which the animal swims.

The best-known British form, *Doliolum tritonis*, has been captured on occasions in thousands off our N.W. coast, between the Hebrides and the Faroes. This species, and the closely allied *D. nationalis*, have also been found in the English Channel and off the S.W. coast of Ireland.

Sub-order II. HEMIMYARIA.

Free-swimming pelagic forms, which exhibit alternation of generations in their life-history, and in the sexual condition form colonies. The body is more or less fusiform, with the long axis antero-posterior, and the branchial and atrial apertures nearly terminal. The test is well developed. The musculature of the body-wall is in the form of a series of transversely-running bands, which do not form complete independent rings, as in the CYCLOMYARIA. The branchial and peribranchial cavities form a continuous space in the interior of the body, opening externally by the branchial and atrial apertures, and traversed obliquely from the dorsal and anterior end to the ventral and posterior by a long, narrow, vascular band, which represents the dorsal lamina, the dorsal blood sinus, and the neighbouring part of the dorsal edge of the branchial sac of an ordinary Ascidian. The alimentary canal is placed ventrally. The embryonic development is direct, no tailed larva being formed.

The Salpidæ, the chief family in this sub-order, includes the single genus *Salpa*, which, however, may be divided into two well-marked groups of species—(1) those, such as *S. pinnata*, in which the alimentary canal is stretched out along the ventral surface of the body, and (2) those, such as *S. fusiformis*, in which the alimentary canal forms a compact globular mass, the "nucleus," near the posterior end of the body. About fifteen species altogether are known; they are all pelagic forms, and are found in many seas. Each species occurs in two forms—the solitary asexual *(proles solitaria)* and the aggregated sexual *(proles gregaria)*—which are usually quite unlike one another. The solitary form gives rise, by internal gemmation, to a complex tubular stolon, which contains processes from all the more important organs of the parent body, and which becomes segmented into a series of buds. As the stolon elongates, the buds near the free end, which have become advanced in their development, are set free in groups, the members in which remain attached together by processes of the test, each enclosing a diverticulum from the body-wall, so as to form "chains." Each member of the chain is a *Salpa* of the sexual or aggregated form, and when mature may—either still attached to its neighbours or separated from them—produce one or several embryos, which develop into the solitary *Salpa*. Thus the two forms alternate regularly.

Salpa, like *Doliolum*, is probably only an occasional visitor in our seas, but several species of the genus— *Salpa democratica-mucronata*, *S. runcinata-fusiformis*, and *S. zonaria*—have been found on occasions in the seas of the Hebrides, or cast ashore on our southern and western coasts.

Order III. ASCIDIACEA.

Fixed or free-swimming Simple or Compound Ascidians, which, in the adult, are never provided with a tail, and have no trace of a notochord. The free-swimming forms are colonies, the Simple Ascidians being always fixed. The test is permanent and well developed; as a rule, it increases with the age of the individual. The branchial sac is large and well developed. Its walls are perforated by numerous slits (stigmata) opening into the peribranchial cavity, which communicates with the exterior by the atrial aperture. Many of the forms reproduce by gemmation, and in most of them the sexually produced embryo develops into a tailed larva.

The Ascidiacea includes three groups — the Simple Ascidians, the Compound Ascidians, and the free-swimming colonial *Pyrosoma*.

Sub-order I. ASCIDLE SIMPLICES.

Fixed Ascidians which are solitary and very rarely reproduce by gemmation; if colonies are formed, the members are not buried in a common investing mass, but each has a distinct test of its own. No strict line of demarcation can be drawn between the Simple and Compound Ascidians, and one of the families of the former group, the Clavelinidæ (the Social Ascidians), forms a transition from the typical Simple forms, which never reproduce by gemmation to the Compound forms, which always do.

The Ascidiæ Simplices may be divided into the following families :—

Family I. CLAVELINIDÆ :— Simple Ascidians which reproduce by gemmation to form small colonies, in which each Ascidiozooid has a distinct test, but all are connected by a common blood-system. Buds formed on stolons, which are vascular out-growths from the posterior end of the body containing prolongations from the ectoderm, mesoderm, and endoderm of the Ascidiozooid. Branchial sac not folded; internal longitudinal bars usually absent; stigmata straight; tentacles simple.

This family contains three chief genera—*Ecteinascidia*, with internal longitudinal bars in branchial sac; *Clavelina*, with intestine extending behind branchial sac; and *Perophora*, with intestine alongside branchial sac. *Clavelina lepadiformis* and *Perophora listeri* are common British species found at a few fathoms depth off various parts of our coast. Both occur round the south end of the Isle of Man.

Family II. ASCIDIIDÆ :—Solitary fixed Ascidians with gelatinous test : branchial aperture usually 8-lobed, atrial aperture usually 6-lobed. Branchial sac not folded; internal longitudinal bars usually present; stigmata straight or curved ; tentacles simple.

This family contains, along with several other genera,

our typical form *Ascidia*, of which there are many species in British seas, widely distributed round our coasts. Two other common British forms, belonging to this family, are *Ciona intestinalis*, with a very soft, pale-green test and languets in place of a dorsal lamina, and *Corella parallelogramma*, in which the stigmata of the branchial sac are curved to form beautiful spirals.

Family III.   CYNTHIIDÆ :— Solitary fixed Ascidians, usually with leathery test ; branchial and atrial apertures both 4-lobed   Branchial sac longitudinally folded ; stigmata straight ; tentacles simple or compound.

This is the largest family of Simple Ascidians, and contains a number of genera, about six of which are British.   *Styela* has simple tentacles, and not more than 4-folds on each side of the branchial sac.   A very common species all round our coasts, between tide marks, is the little red *Styela* (or *Styelopsis*) *grossularia*.   In *Cynthia* the tentacles are compound, and there are more than 4 folds, usually 7 or 8, on each side of the branchial sac. The curious little *Forbesella tessellata*, from deep water in the Irish Sea, is in some respects intermediate between *Styela* and *Cynthia*.

Family IV.   MOLGULIDÆ :— Solitary Ascidians, often not fixed ; branchial aperture 6-lobed, atrial 4-lobed.   Test usually encrusted with sand.   Branchial sac longitudinally folded ; stigmata more or less curved, usually arranged in spirals ; tentacles compound.

Several species of *Molgula*, all looking when dredged like little sandy balls, and one of *Eugyra (E. glutinans)*, are common at a few fathoms depth round most parts of our coasts.

Sub-order II. ASCIDIÆ COMPOSITÆ.

Fixed Ascidians which reproduce by gemmation, so as to form colonies, in which the Ascidiozooids are buried in a common investing mass, and have no separate tests. This is probably a somewhat artificial assemblage, formed of two or three groups of Ascidians which produce colonies in which the Ascidiozooids are so intimately united that they possess a common test or investing mass. This is the only character which distinguishes them from the Clavelinidæ, but the property of reproducing by gemmation separates them from the rest of the Ascidiæ Simplices. The Ascidiæ Compositæ may be divided into the following families :—

Family I. DISTOMIDÆ :—Ascidiozooids divided into two regions, thorax and abdomen; testes numerous; vas deferens not spirally coiled. The chief genera are—*Distoma*, *Distaplia*, *Colella*, the last forming a pedunculated colony, in which the Ascidiozooids develop incubatory pouches, opening from the peribranchial cavity, in which the embryos undergo their development.

Family II. CŒLOCORMIDÆ :—Colony not fixed, having a large axial cavity with a terminal aperture. Branchial apertures 5-lobed. This includes one species, *Cœlocormus huxleyi*, which is a transition form between the ordinary Compound Ascidians (*e.g.*, Distomidæ) and the Ascidiæ Luciæ (*Pyrosoma*).

Family III. DIDEMNIDÆ :—Colony usually thin and incrusting. Test containing stellate calcareous spicules. Testis single, large; vas deferens spirally coiled. The chief genera are—*Didemnum*, in which the colony is thick and fleshy, and there are only three rows of stigmata on each side of the branchial sac ; and *Leptoclinum*, in which the colony is thin and incrusting, and there are four rows of stigmata on each side of the branchial sac. Colonies of *Leptoclinum*, forming thin, white, grey, or yellow crusts, under stones at low water, are amongst the commonest of British Compound Ascidians.

Family IV. DIPLOSOMIDÆ :—Test reduced in amount, rarely containing spicules. Vas deferens not spirally coiled. In *Diplosoma*, the most important genus, the larva is gemmiparous. This is a common British form, especially on *Zostera* beds, and amongst sea-weeds.

Family V. POLYCLINIDÆ :—Ascidiozooids divided into three regions, thorax, abdomen, and post-abdomen. Testes numerous ; vas deferens not spirally coiled. The chief genera are—*Pharyngodictyon*, with stigmata absent or modified, one species : the only Compound Ascidian known from a depth of 1000 fathoms ; *Polyclinum*, with a smooth-walled stomach ; *Aplidium*, with the stomach-wall longitudinally folded ; and *Amaroucium*, in which the Ascidiozooid has a long post-abdomen and a large atrial languet. The last three genera contain many common British species.

Family VI. BOTRYLLIDÆ :—Ascidiozooids having the intestine and reproductive organs alongside the branchial sac. Dorsal lamina present ; internal longitudinal bars present in branchial sac. The chief genera are—*Botryllus*, with simple stellate systems, and *Botrylloides*,* with elongated or ramified systems. There are many species of both these genera, which form brilliantly coloured fleshy crusts under stones and on sea-weed at low tide. They are amongst the commonest and the most beautiful of British Ascidians.

* It is intended that a future L.M.B.C. Memoir will deal with *Botrylloides* as a type of the Compound Ascidians.

Family VII. POLYSTYELIDÆ:— Ascidiozooids not grouped in systems. Branchial and atrial apertures 4-lobed. Branchial sac may be folded; internal longitudinal bars present. The chief genera are—*Thylacium*, with ascidiozooids projecting above general surface of colony; *Goodsiria*, with ascidiozooids completely imbedded in investing mass; and *Chorizocormus*, with ascidiozooids united in little groups, which are connected by stolons. The last genus contains one species, *Ch. reticulatus*, a transition form between the other Polystyelidæ and the Cynthiidæ among Simple Ascidians. *Thylacium* is the only British form.

Sub-order III. ASCIDIÆ LUCIÆ.

Free-swimming pelagic colonies having the form of a hollow cylinder closed at one end. The ascidiozooids forming the colony are imbedded in the common test in such a manner that the branchial apertures open on the outer surface, and the atrial apertures on the inner surface next to the central cavity of the colony. The ascidiozooids are produced by gemmation from a rudimentary larva (the cyathozooid) developed sexually.

This sub-order includes a single family, the PYROSOMIDÆ, containing one well-marked genus, *Pyrosoma*, with several species. They are found swimming near the surface of the sea, chiefly in tropical latitudes, and are brilliantly phosphorescent. A fully developed colony may be from an inch or two to upwards of four feet in length. *Pyrosoma* does not occur in British seas.

49

EXPLANATION OF PLATES.

PLATE I.

Large specimen of *Ascidia mentula*, from the right side, natural size, from life.

Length 11·5 cm.   Breadth 6 cm.   Thickness 3 cm.

A, shows the sensory edge of the branchial aperture with lobes and coloured lines (the " ocelli "), enlarged.

PLATE II.

Reference Letters in Plates II., III., and IV.

a. anus.
At. atrial aperture.
at.l. atrial lobe.
bl c. bladder cell.
bl.s. blood sinus.
Br. branchial aperture.
br.car branchio-cardiac vessel.
br.ao. branchial (ventral) vessel.
Br.s. branchial sac.
br.visc. branchio-visceral vessel.
card.visc. cardio-visceral vessel.
c.c. common cloaca.
c.d. connecting duct.
cl. cloaca.
con. connective.
c t.c. connective tissue cell.
d.ao. dorsal vessel.
d.bl.s. dorsal blood sinus.
d.l. dorsal lamina.
d.n. dorsal nerve cord.
d.t. dorsal tubercle.
ec. ectoderm.
en. endoderm.
end. endostyle.
ep.c. epicardial tube.

g.d. genital ducts.
gl. pyloric gland.
gl.d. duct of neural gland.
gon. gonads.
h. heart.
h m. horizontal membrane.
hyp. hypophysial (neural) gland
hyp d. hypophysial duct.
i.i'. intestine
i.l. internal longitudinal bars.
i.v. interstigmatic vessels.
m. body-wall, or mantle.
m.b. muscle bands.
mes c. mesoblast cell.
m.f. muscle fibres.
n. nerves.
n.g. nerve ganglion.
n.gl. neural gland.
œs. œsophagus.
ov. ovary.
p.br. peribranchial cavity.
p.br.z. prebranchial zone.
p c. pericardium.
p.p'. papillæ.
pp.b. peripharyngeal bands.

50

r. rectum.	*tn*. tentacle.
ren. renal vesicles.	*t.k*. terminal knobs on vessels.
sg. stigmata.	*tr*. transverse vessels.
sp. testis.	*t.v*. test vessel.
sph. sphincter.	*ty*. typhlosole.
st. stomach.	*v*. vessel.
t. test.	*v.app*. vascular appendage.
t.c. test cells.	*v.bl.s*. ventral blood sinus.

Fig. 1. Diagram of the outside of *Ascidia*, from right side.

Fig. 2. Transverse section through the atrial aperture to show arrangement of internal cavities and organs. (To avoid complication the internal longitudinal bars are not represented in the branchial sac.)

Fig. 3. A mesh of the branchial sac, diagrammatic.
A. From the inner surface. B. In section.

Fig. 4. Diagrammatic dissection to show the structure of *Ascidia* (compare with fig. 2).

Fig. 5. Section through test and mantle (body-wall) to show the relations of the ectoderm and mesoderm to the test.

Fig. 6. Sagittal section through antero-dorsal part of body to show relations of nerve ganglion, neural gland, &c.

Fig. 7. Dorsal front of pharynx from inside, to show dorsal tubercle, tentacles, and neighbouring parts. × 50.

PLATE III.

Figs. 1 to 4 show typical examples of four important families of Compound Ascidians, natural size.

Fig. 1 is a Distomid colony (*Colella*).

Fig. 2 is a Didemnid colony (*Leptoclinum*).

Fig. 3 is a Polyclinid colony (*Pharyngodictyon*).

Fig. 4 is a Botryllid colony (*Botryllus*).

Fig. 5 shows the union of two Ascidiozooids, in a Compound Ascidian colony, to form a common cloaca (*c.c.*) in the common test.

Figs. 6 to 8 show the three forms of body found amongst the Ascidiozooids of Compound Ascidians.

Fig. 6 is from a Botryllid colony.

Fig. 7 is from a Distomid colony.

Fig. 8 is from a Polyclinid colony (all from right side, magnified).

Fig. 9. Section through outer part of test of *Ascidia* to show " vessels " and " bladder " cells. × 50.

Fig. 10. Diagram of *Ascidia* to show the arrangement of the blood system, from left side, in correct morphological position for comparison with Vertebrates.

PLATE IV

Fig. 1. Part of branchial sac of *Ascidia mentula*, from the outside. × 50.

Fig. 2. Part of branchial sac of *Ascidia mentula* from the inside. × 50.

Fig. 3. Small part of last. × 300.

Fig. 4. Part of the dorsal lamina. × 40.

Fig. 5. Transverse section of the endostyle. × 200.

Fig. 6. Blood corpuscles of *Ascidia*.

Fig. 7. Part of section of wall of intestine showing refringent organ, spermatic tubules, &c.

Fig. 8. Ocellus from the branchial aperture, in section.

Fig. 9. The heart and pericardium, in side view.

Fig. 10. The heart and pericardium, in section.

Fig. 11. Group of renal vesicles. × 50.

Fig. 12. Part of a renal vesicle, more highly magnified.

Fig. 13. Two cells from wall of renal vesicle.

Fig. 14. Concretions from renal vesicles. × 200.

Fig. 15. Bunch of spermatic tubules from testis. × 50.

PLATE V.

REFERENCE LETTERS.—*ar.* archenteron, *at.* atrial involution, *au.* otolith, *bc.* blastocoele, *bp.* blastopore, *ch.* notochord, *ep.* epiblast, *f.* tail fin, *foll.* follicle cell, *hy.* hypoblast, *i.* intestine, *m.* mouth, *mes.* mesoblast, *musc.* muscle cells, *mb.* mesoblast layer, *mc.* mesoblast cell, *n.* nucleus of ovum, *nc.* neural canal. *n.e.c.* neurenteric canal, *n.ch.* notochord, *n.v.* cerebral vesicle, *oc.* cerebral eye, *p.* protoplasm of ovum, *s.* spermatozoon, *t.c.* "testa" cells of ovum.

Fig. 1. Mature ovum and spermatozoon of *Ascidia.*

Fig. 2. A segmentation stage, in section, to show blastula.

Fig. 3. Early gastrula stage.

Fig. 4. Later gastrula stage.

   In these and some of the other figures, the clear cells indicate the notochord, the hypoblast cells have small circles, and the nerve cells are marked with oblique lines, while the mesoblast cells are cross hatched.

Fig. 5. A later embryo, showing rudiments of notochord and nervous system.

Fig. 6. Embryo showing body and tail, &c.

Fig. 7. Transverse section of the tail of larva.

Fig. 7a. Transverse section of the body of embryo.

Fig. 8. Embryo ready to be hatched.

Fig. 9. Free-swimming tailed larva.

Fig. 10. The metamorphosis—larva attached to a stone.

Fig. 11. Tail and nervous system of larva degenerating.

Fig. 12. Further degeneration and metamorphosis of larva into

Fig. 13. the young fixed Ascidian.

ASCIDIA.

S B lith

W A H del

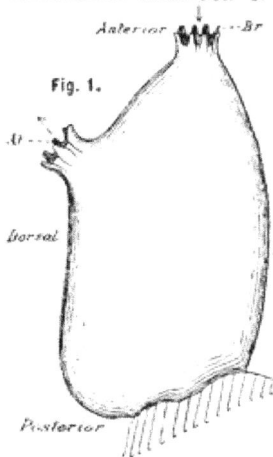

Fig. 1.

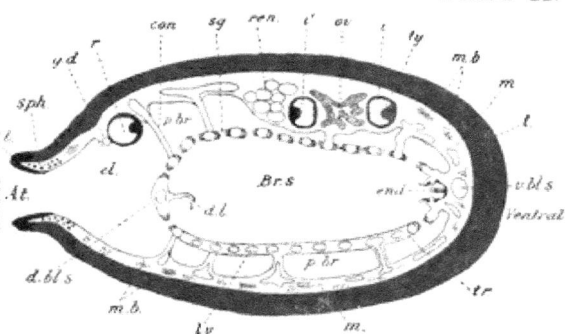

Fig. 2.

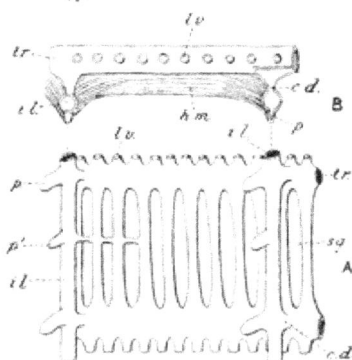

Fig. 3.

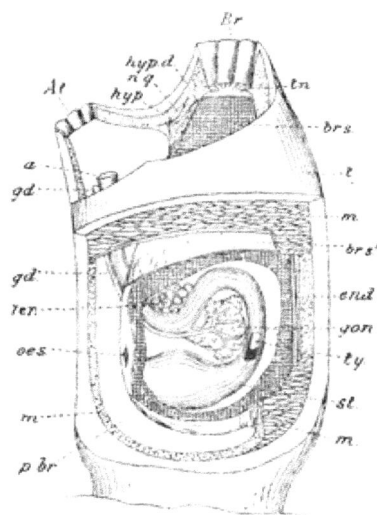

Fig. 4.

Fig. 5.

Fig. 6.

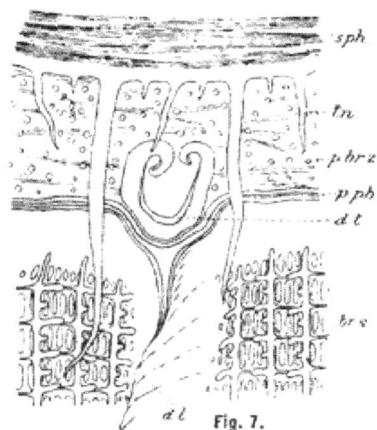

Fig. 7.

W.A.H. del.   ASCIDIA.

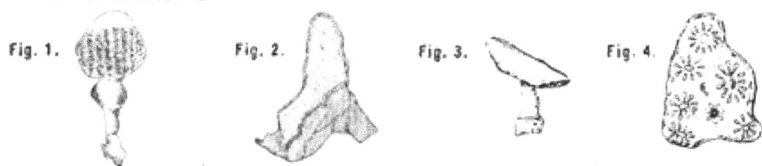

Fig. 1.   Fig. 2.   Fig. 3.   Fig. 4.

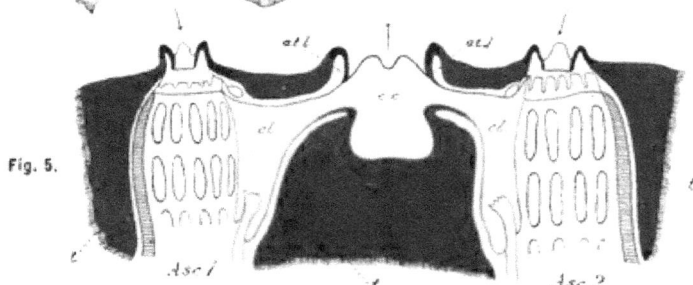

Fig. 5.

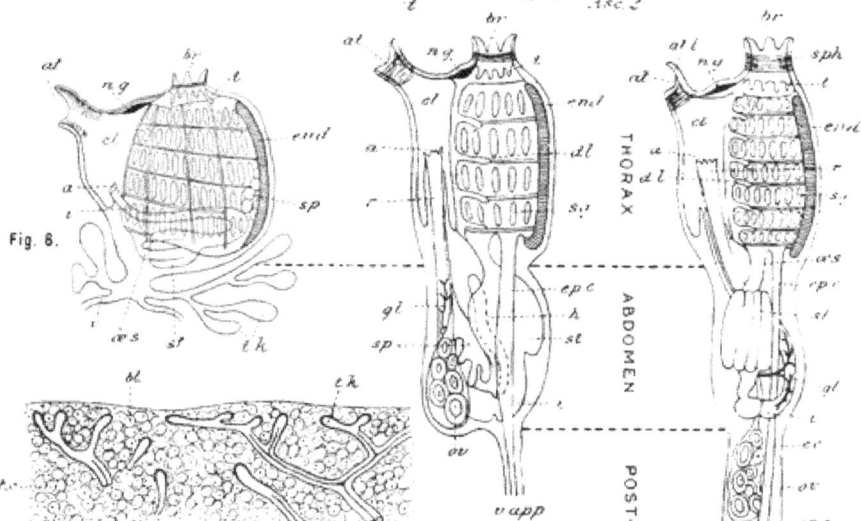

Fig. 8.

Fig. 7.

Fig. 8.

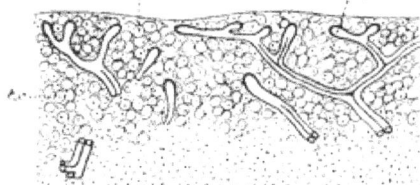

Fig. 9.

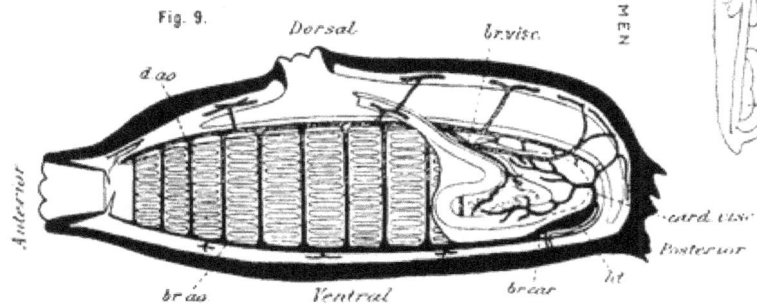

Fig. 10.

W.A.H. del.

ASCIDIA.

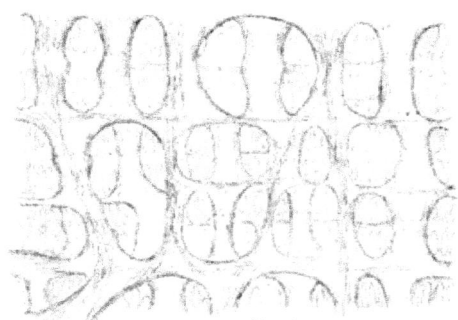

Fig.1.

Fig.2.

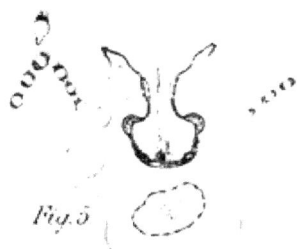

Fig.5.

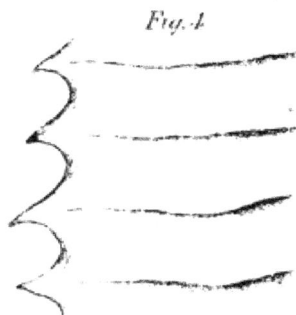

Fig.4.

Fig.3.

Fig.6.

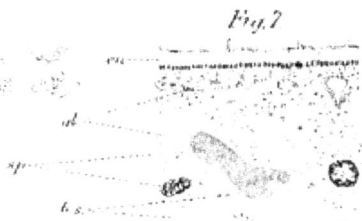

Fig.7.

Fig.8.

Fig.15.

Fig.9.

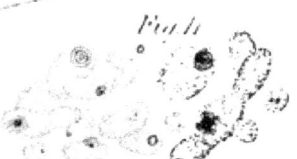

Fig.11.

Fig.14.

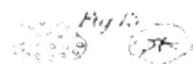

Fig.13.

Fig.12.

Fig.10.

W.A.H.del ASCIDIA. S.B.lith.

Fig. 1.

Fig. 2.

Fig. 3.

Fig. 4.

Fig. 5.

Fig. 6.

Fig. 7.

Fig. 7a.

Fig. 8.

Fig 9.

Fig. 10.

Fig. 11.

Fig. 12.

Fig. 13.

W.A.H. del.

ASCIDIA.

www.ingramcontent.com/pod-product-compliance
Lightning Source LLC
Chambersburg PA
CBHW022000190326
41519CB00010B/1342